Mifflin
Harcourt

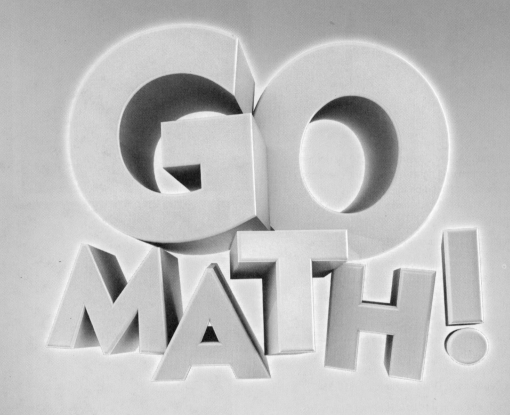

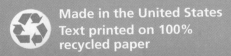

Made in the United States
Text printed on 100%
recycled paper

Houghton
Mifflin
Harcourt

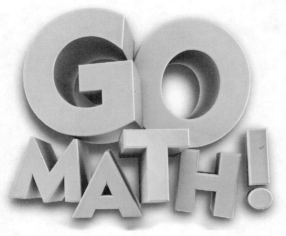

ISBN 978-0-544-34188-3

22 0877 21

4500827631   C D E F G

Dear Students and Families,

Welcome to **Go Math!**, Grade 1! In this exciting mathematics program, there are hands-on activities to do and real-world problems to solve. Best of all, you will write your ideas and answers right in your book. In **Go Math!**, writing and drawing on the pages helps you think deeply about what you are learning, and you will really understand math!

By the way, all of the pages in your **Go Math!** book are made using recycled paper. We wanted you to know that you can Go Green with **Go Math!**

Sincerely,

The Authors

Made in the United States
Text printed on 100% recycled paper

GO MATH!

Authors

Juli K. Dixon, Ph.D.
Professor, Mathematics Education
University of Central Florida
Orlando, Florida

Edward B. Burger, Ph.D.
President, Southwestern University
Georgetown, Texas

Steven J. Leinwand
Principal Research Analyst
American Institutes for
    Research (AIR)
Washington, D.C.

Matthew R. Larson, Ph.D.
K-12 Curriculum Specialist for
    Mathematics
Lincoln Public Schools
Lincoln, Nebraska

Martha E. Sandoval-Martinez
Math Instructor
El Camino College
Torrance, California

Contributor

Rena Petrello
Professor, Mathematics
Moorpark College
Moorpark, California

English Language Learners Consultant

Elizabeth Jiménez
CEO, GEMAS Consulting
Professional Expert on English
    Learner Education
Bilingual Education and
    Dual Language
Pomona, California

Operations and Algebraic Thinking

  Critical Area Developing understanding of addition, subtraction, and strategies for addition and subtraction within 20.

5 Addition and Subtraction Relationships 251

COMMON CORE STATE STANDARDS

1.OA Operations and Algebraic Thinking
Cluster A Represent and solve problems involving addition and subtraction.
1.OA.A.1
Cluster C Add and subtract within 20.
1.OA.C.6
Cluster D Work with addition and subtraction equations.
1.OA.D.7
1.OA.D.8

GO DIGITAL

Go online! Your math lessons are interactive. Use *iTools*, Animated Math Models, the Multimedia *eGlossary*, and more.

Chapter 5 Overview

In this chapter, you will explore and discover answers to the following **Essential Questions**:

• How can relating addition and subtraction help you to learn and understand facts within 20?

• How do addition and subtraction undo each other?

• What is the relationship between related facts?

• How can you find unknown numbers in related facts?

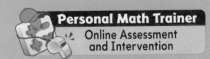

Personal Math Trainer
Online Assessment and Intervention

v

CRITICAL AREA REVIEW PROJECT MAKE A MATH FACTS STRATEGIES BOOK: *www.thinkcentral.com*

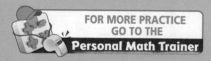

FOR MORE PRACTICE
GO TO THE
Personal Math Trainer

Practice and Homework

Lesson Check and
Spiral Review in
every lesson

Chapter 5

Addition and Subtraction Relationships

Curious About Math with
Curious George

Children tap the Liberty Bell 4 times. Then they tap it 9 more times. How many times do the children tap the bell?

Name _____

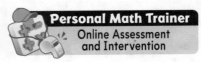
Personal Math Trainer
Online Assessment
and Intervention

Add in Any Order

Use 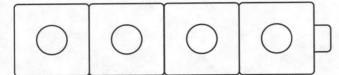. Color to match.
Write each sum. (K.OA.A.1)

1.

$1 + 3 =$ _____

$3 + 1 =$ _____

Count On

Count on to add. Write each sum. (K.OA.A.5)

2. $6 + 3 =$ ____ | 3. $7 + 1 =$ ____ | 4. $8 + 2 =$ ____

Count Back

Count back to subtract. Write each difference. (K.OA.A.5)

5. $11 - 2 =$ ____ | 6. $8 - 3 =$ ____ | 7. $9 - 1 =$ ____

This page checks understanding of important skills needed
for success in Chapter 5.

Name _____

Vocabulary Builder

Review Words
add
addition fact
difference
subtract
subtraction fact
sum

Visualize It

Sort the review words from the box.

Addition Words   Subtraction Words

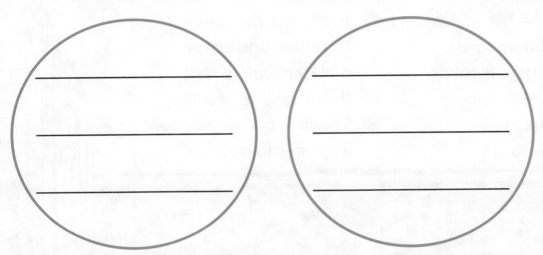

Understand Vocabulary

Follow the directions.

1. Write an addition fact.

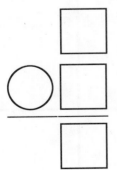

2. What is the sum?

____

3. Write a subtraction fact.

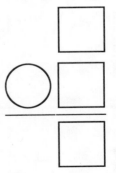

4. What is the difference?

____

Game Add to Subtract Bingo

Materials • 16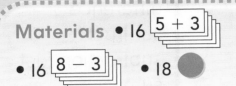

• 16 $8 - 3$ • 18 ●

Play with a partner.

Each player picks ● or ◐.

1. Mix the addition cards. Each player gets 8 cards. Show your cards faceup.

2. Put the subtraction cards in a pile facedown.

3. Take a subtraction card. Do you have the addition fact that helps you subtract?

4. If so, put the cards together and cover a space with a ●. If not, you lose a turn.

5. The first player to cover 3 spaces in a row wins.

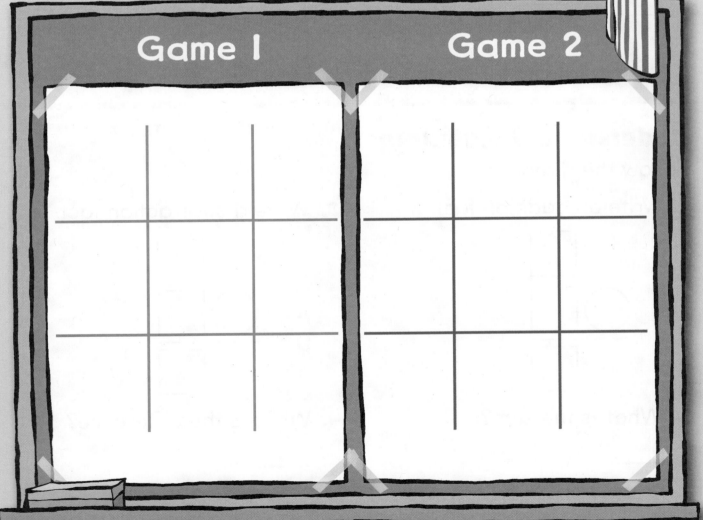

Chapter 5 Vocabulary

add

sumar

1

addend

sumando

2

addition sentence

enunciado de suma

3

difference

diferencia

13

related facts

operaciones relacionadas

47

subtract

restar

52

subtraction sentence

enunciado de resta

53

sum

suma o total

54

5   +   3   =   8

↑
addends

$3 + 2 = 5$

$9 - 4 = 5$

The **difference** is 5.

4   +   2   =   6

is an **addition sentence.**

$5 - 2 = 3$

Related facts have the same parts and wholes.

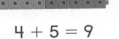

$4 + 5 = 9$ $9 - 5 = 4$

$5 + 4 = 9$ $9 - 4 = 5$

2 plus 1 is equal to 3.

The **sum** is 3.

$9 - 5 = 4$

is a **subtraction sentence.**

Make a Match

Materials
2 sets of word cards

How to Play
Play with a partner.

1. Mix the cards. Give 5 cards to each player. Put the rest in a pile facedown.

2. Ask your partner for a word to match a word in your hand.

   - If the partner has the word, he or she gives the card to you. Place the pair of word cards to the side.

   - If the partner does not have the word, he or she says, "Make a match." Take a card from the pile. If that card matches a word in your hand, set the pair of word cards to the side.

3. Take turns.

4. When one partner has matched all his or her word cards, that player wins. When there are no more word cards in the pile, the player with the most matches wins.

The Write Way

Reflect

Choose one idea. Draw and write about it.

- Max wants to know if the following is correct.

$$3 + 5 = 6 + 2$$

Draw and write to tell how you know.

- Tell three things you know about adding and subtracting numbers.

Name _____

Problem Solving • Add or Subtract

Essential Question How can making a model help you solve a problem?

Common Core **Operations and Algebraic Thinking—1.OA.A.1**
MATHEMATICAL PRACTICES
MP1, MP2, MP4

There are 16 turtles on the beach. Some swim away. Now there are 9 turtles on the beach. How many turtles swim away?

 **Unlock the Problem** Real World

What do I need to find?

how many ~~turtles~~ swim away

What information do I need to use?

16 turtles

? swim away

9 turtles now on the beach

Show how to solve the problem.

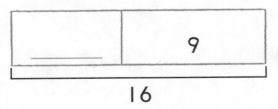

| 9 |
16

16 turtles _____ swim away   9 turtles now on the beach

HOME CONNECTION • Your child made a model to visualize the problem. The model helps your child see what part of the problem to find.

Chapter 5

Try Another Problem

Make a model to solve.
Use to help you.

- What do I need to find?
- What information do I need to use?

1. There are 4 rabbits in the garden.
Some more rabbits come.
Now there are 12 rabbits. How many rabbits come to the garden?

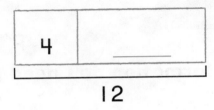

4	
4 rabbits   _____ rabbits come   12 rabbits in the garden

2. There are 14 birds in a tree.
Some birds fly away. There are 9 birds still in the tree.
How many birds fly away?

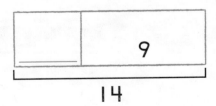

14 birds   _____ birds fly away   9 birds still in the tree

Math Talk

MATHEMATICAL PRACTICES 4

Model Explain how to find the missing number.

Name _____

Make a model to solve.

3. There are 20 ducks in the pond.
Then 10 ducks swim away.
How many ducks are
still in the pond?

10	_____
20	

20 ducks   10 swim away   _____ ducks still in the pond

4. **THINK SMARTER** 3 eagles land
in the trees. Now
12 eagles are in the
trees. How many
eagles were in the
trees to start?

_____	3
12	

_____ eagles   3 eagles land   12 eagles in the trees

5. 8 squirrels are in the park.
Some more squirrels come.
Now there are 16 squirrels.
How many squirrels come
to the park?

8	_____
16	

8 squirrels   _____ squirrels come   16 squirrels in the park

On Your Own   WRITE Math

MATHEMATICAL PRACTICE ② Represent a Problem

Solve. Draw or write to show your work.

6. Liz picks 15 flowers.
7 are pink. The rest
are yellow. How
many are yellow?

_____ yellow flowers

7. Cindy has 14 sand dollars.
She has the same number
of large and small sand
dollars. Write a number
sentence about
the sand dollars.

___ ◯ ___ ◯ ___

8. **GO DEEPER** Sam has 3 more books
than Ed. Sam has 8 books. How
many books does Ed have?

_____ books

9. **THINK SMARTER** There are 7 eggs
in a nest. Some eggs hatch.
Now there are 5 left. How
many eggs hatch?

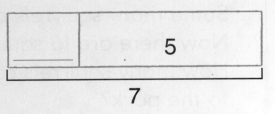

7 eggs   _____ hatch   5 eggs left

TAKE HOME ACTIVITY • Ask your child to look at Exercise 7
and use the number 18 as the total number of sand dollars.
Then have your child write a number sentence.

© Houghton Mifflin Harcourt Publishing Company • Image Credits: (c) ©Siede Preis/PhotoDisc/Getty Images (t) ©America/Alamy

Add or Subtract

Common Core

COMMON CORE STANDARD—1.OA.A.1
Represent and solve problems involving
addition and subtraction.

Make a model to solve.

1. Stan has 12 stickers.
Some stickers are new.
4 stickers are old.
How many stickers are new?

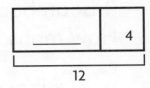

_____ new stickers

2. Liz has 9 toy bears.
Then she buys some more.
Now she has 15 toy bears.
How many toy bears did she buy?

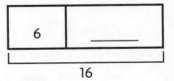

_____ toy bears

3. Eric buys 6 books.
Now he has 16 books.
How many books did he
have to start?

_____ books

4. WRITE Math   Write an
addition story problem.
Have a classmate solve
your problem.

_____

_____

_____

_____

Lesson Check (1.OA.A.1)

Use the model to solve.

1. Arlo has 17 bean bag animals.
   Some are fuzzy.
   9 bean bag animals are not fuzzy.
   How many animals are fuzzy?

   _____ fuzzy animals

	9
_____	

17

Spiral Review (1.OA.A.1, 1.OA.C.5)

2. Count back.
   Write the difference.

$$ \underline{\quad} = 11 - 3 $$

3. Use 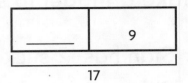. Color to show how to make ten.
   Complete the addition sentence.

   ☐☐☐☐☐☐☐☐☐☐

   $$ 10 = \underline{\quad} + \underline{\quad} $$

FOR MORE PRACTICE
GO TO THE
Personal Math Trainer

Name _____

Record Related Facts

Essential Question How do related facts help you find missing numbers?

Common Core **Operations and Algebraic Thinking—1.OA.C.6** *Also 1.OA.D.8*
MATHEMATICAL PRACTICES
MP1, MP5, MP7, MP8

Listen and Draw · Real World · Hands On

Listen to the problem.
Model with �_▪ ▪_ or an *i*Tool. Draw your model.
Write the number sentence.

___ + ___ = ___   ___ − ___ = ___

Math Talk
MATHEMATICAL PRACTICES 5

Use Tools Explain how your model helps you write your number sentence.

FOR THE TEACHER • Read the following problem for the left box. Colin has 7 crackers. He gets 1 more cracker. How many crackers does Colin have now? Then read the following for the right box. Colin has 8 crackers. He gives one to Jacob. How many crackers does Colin have now?

Chapter 5

Model and Draw

How can one model help you write four **related facts**?

4 + 5 = 9

5 + 4 = 9

9 − 5 = 4

9 − 4 = 5

Share and Show   MATH BOARD

Use . Add or subtract.
Complete the related facts.

1. 8 + ☐ = 15   15 − 7 = ☐

    7 + 8 = ☐   ☐ − ☐ = ☐

2. ☐ + 9 = 14   14 − ☐ = 5

    9 + 5 = ☐   ☐ − ☐ = ☐

3. 7 + ☐ = 13   13 − 6 = ☐

    6 + 7 = ☐   ☐ − ☐ = ☐

Name _____

On Your Own

MATHEMATICAL PRACTICE ①
Analyze Relationships

Use . Add or subtract. Complete the related facts.

4. $\boxed{} + 8 = 13$ $13 - \boxed{} = 5$

  $8 + 5 = \boxed{}$ $\boxed{} - \boxed{} = \boxed{}$

5. $\boxed{} + 8 = 17$ $17 - \boxed{} = 9$

  $8 + 9 = \boxed{}$ $\boxed{} - \boxed{} = \boxed{}$

6. $9 + \boxed{} = 15$ $\boxed{} - 6 = 9$

  $6 + \boxed{} = 15$ $\boxed{} - \boxed{} = \boxed{}$

7. **THINK SMARTER** Circle the number sentence that has a mistake. Correct it to complete the related facts.

  $7 + 9 = 16$

  $16 + 9 = 7$

  $9 + 7 = 16$

  $16 - 7 = 9$

____ ◯ ____ ◯ ____

Problem Solving • Applications  WRITE Math

8. **GO DEEPER** Choose three numbers to make
related facts. Choose numbers between
0 and 18. Write your numbers.
Write the related facts.

9. **THINK SMARTER** Which fact is a related fact?

$$6 + 3 = 9 \qquad 9 - 3 = 6$$
$$3 + 6 = 9 \qquad ?$$

○ $6 + 9 = 15$
○ $9 + 3 = 12$
○ $9 - 6 = 3$
○ $6 - 3 = 3$

 TAKE HOME ACTIVITY • Write an addition fact.
Ask your child to write three other related facts.

Record Related Facts

Common Core

COMMON CORE STANDARD—1.OA.C.6
Add and subtract within 20.

Use 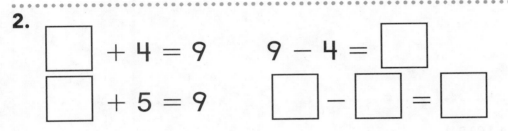 . Add or subtract. Complete
the related facts.

1. $4 + \boxed{} = 12$ $\boxed{} - 8 = 4$

    $8 + 4 = \boxed{}$ $\boxed{} - \boxed{} = \boxed{}$

2. $\boxed{} + 4 = 9$ $9 - 4 = \boxed{}$

    $\boxed{} + 5 = 9$ $\boxed{} - \boxed{} = \boxed{}$

Problem Solving Real World

Choose a way to solve.
Write or draw to explain.

3. There are 16 apples on
   the tree. No apples fall off.
   How many apples are still on the tree?

   _____ apples

4. WRITE Math Write four
   related facts. Use pictures
   to show how the number
   sentences are related.

_____

_____

Lesson Check (1.OA.C.6)

1. Write a related fact.

$$7 + 4 = 11$$
$$4 + 7 = 11$$

$$11 - 7 = 4$$
$$\boxed{} - \boxed{} = \boxed{}$$

- -

Spiral Review (1.OA.B.4, 1.OA.D.8)

2. Complete the subtraction sentence.

$$6 - 6 = \underline{\qquad}$$

- -

3. Write an addition sentence that helps you solve $15 - 9$.

$$\underline{\qquad} + \underline{\qquad} = \underline{\qquad}$$

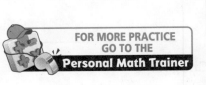

Name _____

Identify Related Facts

Essential Question How do you know if
addition and subtraction facts are related?

 Common Core **Operations and Algebraic Thinking—1.OA.C.6** *Also 1.OA.D.8*
MATHEMATICAL PRACTICES
MP4, MP7, MP8

Listen and Draw

Use ▪▪ ▪▪ to show 4 + 9 = 13.
Draw ▪▪ ▪▪ to show a related subtraction fact.
Write the subtraction sentence.

Math Talk

MATHEMATICAL PRACTICES 7

Look for Structure
Why is your subtraction
sentence related to
4 + 9 = 13?

FOR THE TEACHER • Have children use cubes to
show 4 + 9 = 13. Then have them use cubes to
show the related subtraction sentence, draw the
cubes, and write the related sentence.

Chapter 5

Model and Draw

Use the pictures. What two facts can you write?

_3_ (+) _9_ (=) _12_

_12_ (−) _9_ (=) _3_

These are related facts. If you know one of these facts, you also know the other fact.

Share and Show   MATH BOARD

Add and subtract.
Circle the related facts.

1. $6 + 4 =$ ___
   $10 - 4 =$ ___

2. ___ $= 9 + 8$
   ___ $= 17 - 8$

3. $9 + 5 =$ ___
   $9 - 5 =$ ___

4. $8 + 7 =$ ___
   $15 - 7 =$ ___

5. ___ $= 9 + 2$
   ___ $= 9 - 2$

6. $6 + 3 =$ ___
   $12 - 3 =$ ___

7. $4 + 8 =$ ___
   $12 - 8 =$ ___

8. ___ $= 7 + 6$
   ___ $= 13 - 6$

9. $9 + 9 =$ ___
   $18 - 9 =$ ___

Name _____

10. **MATHEMATICAL PRACTICE 7** **Identify Relationships** Add and subtract.
Color the leaves 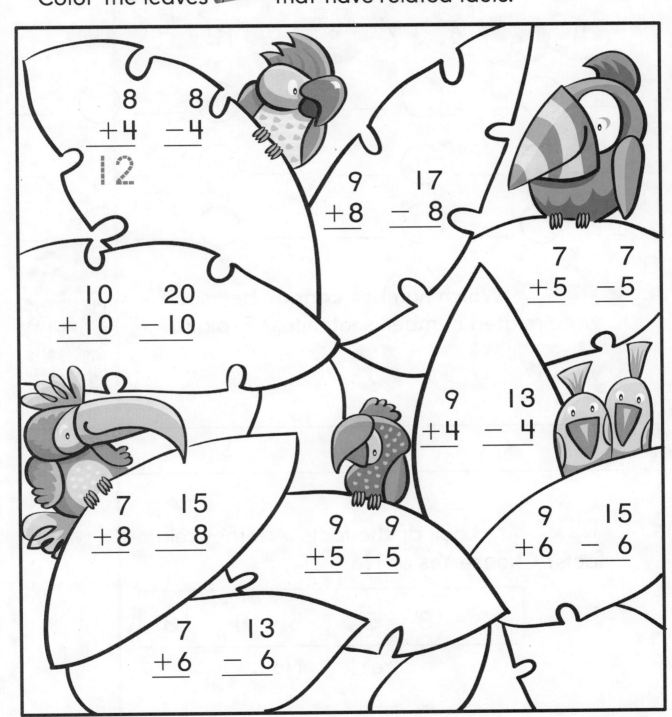 that have related facts.

8 8
+4 −4
——
12

9 17
+8 − 8

7 7
+5 −5

10 20
+10 −10

9 13
+4 − 4

7 15
+8 − 8

9 9
+5 −5

9 15
+6 − 6

7 13
+6 − 6

Problem Solving • Applications   WRITE Math

GO DEEPER Use the numbers to write related addition and subtraction sentences.

4   5   6   7   8   9   12   13   14

11. ___ ◯ ___ ◯ ___   ___ ◯ ___ ◯ ___

12. ___ ◯ ___ ◯ ___   ___ ◯ ___ ◯ ___

13. ___ ◯ ___ ◯ ___   ___ ◯ ___ ◯ ___

14. **THINK SMARTER** Which number **cannot** be used to write related number sentences? Explain.

_____

_____

15. **THINK SMARTER** Look at the facts. Are they related facts? Choose Yes or No.

$13 - 8 = 5$	$5 + 8 = 13$

Yes   No

 TAKE HOME ACTIVITY • Write 7, 9, 16, +, −, and = on separate slips of paper. Have your child use the slips of paper to show related facts.

Name _____

Identify Related Facts

Common Core **COMMON CORE STANDARD—1.OA.C.6**
Add and subtract within 20.

Add and subtract.
Circle the related facts.

1. $5 + 6 =$ ___
    $11 - 6 =$ ___

2. $4 + 9 =$ ___
    $9 - 4 =$ ___

3. $4 + 7 =$ ___
    $11 - 7 =$ ___

4. $9 + 8 =$ ___
    $17 - 8 =$ ___

5. $5 + 7 =$ ___
    $7 - 5 =$ ___

6. $6 + 8 =$ ___
    $14 - 8 =$ ___

Problem Solving Real World

7. Use the numbers to write related addition and subtraction sentences.

 6   7   8   9   15   16   17

___ $+$ ___ $=$ ___   ___ $-$ ___ $=$ ___

8. **WRITE** Math Use numbers and pictures to show related facts with the numbers 7, 9, and 16.

I. Write a related fact for $7 + 6 = 13$.

___ ◯ ___ = ___

2. Draw lines to match. Subtract to compare.
How many fewer 🪶 than 🐦 are there?

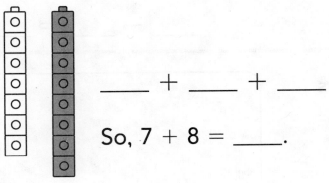

___ − ___ = ___   ___ fewer

3. Use doubles to help you add $7 + 8$.

___ + ___ + ___

So, $7 + 8 =$ ___.

$7 + 8$

FOR MORE PRACTICE
GO TO THE
Personal Math Trainer

Name _____

Use Addition to Check Subtraction

Essential Question How can you use addition to check subtraction?

Common Core Operations and Algebraic Thinking—1.OA.C.6 *Also 1.OA.D.8*
MATHEMATICAL PRACTICES
MP4, MP7, MP8

Listen and Draw (Real World)

Draw and write to solve the problem.

Math Talk MATHEMATICAL PRACTICES 7

Look for Structure Does Erin get all her books back? Use the number sentences to explain how you know.

FOR THE TEACHER • Read the problem. Erin has 11 books. I borrow 4 of them. How many books does Erin still have? Allow children time to solve, using the top workspace. Then read this part of the problem: I give 4 books back to Erin. How many books does Erin have now?

Why can you use addition to check subtraction?

You subtract one part from the whole. The difference is the other part.

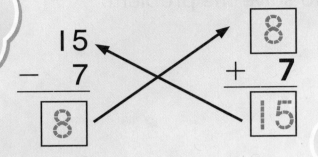

$$\begin{array}{r} 15 \\ -7 \\ \hline \boxed{8} \end{array}$$

$$\begin{array}{r} \boxed{8} \\ +7 \\ \hline \boxed{15} \end{array}$$

When you add the parts, you get the same whole.

Share and Show MATH BOARD

Subtract. Then add to check your answer.

1.
$$\begin{array}{r} 13 \\ -7 \\ \hline \square \end{array}$$
$$\begin{array}{r} \square \\ +7 \\ \hline \square \end{array}$$

2.
$$\begin{array}{r} 14 \\ -5 \\ \hline \square \end{array}$$
$$\begin{array}{r} \square \\ +5 \\ \hline \square \end{array}$$

3.
$$\begin{array}{r} 12 \\ -5 \\ \hline \square \end{array}$$
$$\begin{array}{r} \square \\ +5 \\ \hline \square \end{array}$$

4.
$$\begin{array}{r} 17 \\ -9 \\ \hline \square \end{array}$$
$$\begin{array}{r} \square \\ +9 \\ \hline \square \end{array}$$

Name _____

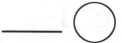  **Look for Structure** Subtract.
Then add to check your answer.

5. $11 - 3 = \boxed{}$

$\boxed{} + 3 = \boxed{}$

6. $13 - 9 = \boxed{}$

$\boxed{} + 9 = \boxed{}$

7.  **THINK SMARTER** Brianna has 13 sand dollars.
Some sand dollars are broken. 5 sand dollars
are not broken. Write number sentences
about the sand dollars.

___ ◯ ___ ◯ ___

___ ◯ ___ ◯ ___

Math
on the
Spot

8. **GO DEEPER** Subtract to solve.
Then add to check your answer.

Liam took 15 balloons to
the party. All but 6 of the
balloons were red. How
many balloons were red?

____ **red balloons**

$\boxed{}$ $\boxed{}$

$-\ \boxed{}$ $+\ \boxed{}$

$\boxed{}$ $\boxed{}$

 TAKE HOME ACTIVITY • Write $11 - 7 = \square$ on a sheet of paper.
Ask your child to find the difference and then write an addition
sentence he or she can use to check the subtraction.

Name _____

 Mid-Chapter Checkpoint

 **Concepts and Skills**

Use ▣ ▣. Add or subtract.
Complete the related facts. (1.OA.C.6)

1. ☐ + 8 = 14   14 − ☐ = 6

    8 + 6 = ☐   ☐ − ☐ = ☐

2. 7 + ☐ = 13   ☐ − 6 = 7

    6 + ☐ = 13   ☐ − ☐ = ☐

Add and subtract. Circle the related facts. (1.OA.C.6)

3. 9 + 3 = ___ | 4. 7 + 8 = ___ | 5. ___ = 6 + 5
    9 − 3 = ___ | 15 − 8 = ___ | ___ = 6 − 5

Personal Math Trainer

6. **THINK SMARTER +** Complete the subtraction.
    Then write an addition sentence to check
    the subtraction. (1.OA.C.6)

    11 − 2 = ☐

    ___ ◯ ___ ◯ ___

Use Addition to Check Subtraction

Common Core **COMMON CORE STANDARD—1.OA.C.6**
Add and subtract within 20.

Subtract. Then add to check your answer.

1. $12 - 4 = \boxed{}$

$\boxed{} + 4 = \boxed{}$

2. $15 - 9 = \boxed{}$

$\boxed{} + 9 = \boxed{}$

Problem Solving Real World

Subtract.
Then add to check your answer.

3. There are 13 grapes in a bowl.
Justin ate some of them.
Now there are only 7 grapes left.
How many grapes did Justin eat?

___ − ___ = ___   ___ grapes.

___ + ___ = ___

4. **WRITE** Math Find $12 - 9$.
Then write or draw
how you can add to
check your answer.

Lesson Check (1.OA.C.6)

1. Subtract. Then add to check your answer.

$$11 - 3 = \boxed{}$$

___ + ___ = ___

2. Subtract. Then add to check your answer.

$$12 - 8 = \boxed{}$$

___ + ___ = ___

Spiral Review (1.OA.A.1, 1.OA.B.3)

3. Jonas picks 10 peaches.
4 peaches are small.
The rest are big.
How many are big?

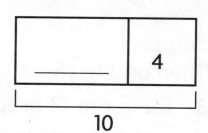

___ big peaches

4. Circle two addends to add first. Write the sum.

FOR MORE PRACTICE GO TO THE Personal Math Trainer

Name _____

Algebra • Unknown Numbers

Essential Question How can you use
a related fact to find an unknown number?

Common
Core

Operations and Algebraic
Thinking—1.OA.D.8 *Also 1.OA.C.6*
MATHEMATICAL PRACTICES
MP1, MP7, MP8

Listen and Draw Real World   Hands On

Listen to the problem. Use ▣▣ to show
the story. Draw to show your work.

FOR THE TEACHER • Read the problem. Calvin
has 7 toy cars that are red. He has some blue toy
cars. He has 10 toy cars. How many blue toy cars
does Calvin have?

Math Talk

MATHEMATICAL PRACTICES

Describe How many toy
cars are blue? Explain how
you got your answer.

Model and Draw

What are the unknown numbers?

$$8 + \boxed{3} = 11$$

$$11 - 8 = \boxed{3}$$

> Use what you know about related facts to find the unknown parts.

Share and Show MATH BOARD

Use ▪▪ ▪▪ to find the unknown numbers.
Write the numbers.

1. $8 + \boxed{} = 15$

   $15 - 8 = \boxed{}$

2. $13 = 9 + \boxed{}$

   $\boxed{} = 13 - 9$

3. $5 + \boxed{} = 14$

   $14 - 5 = \boxed{}$

4. $14 = 6 + \boxed{}$

   $\boxed{} = 14 - 6$

✓ 5. $9 + \boxed{} = 16$

   $16 - 9 = \boxed{}$

✓ 6. $17 = 8 + \boxed{}$

   $\boxed{} = 17 - 8$

280 two hundred eighty

Name _____

HINT
Use a related fact
to help you.

MATHEMATICAL PRACTICE 7 Identify Relationships

Write the unknown numbers.
Use  if you need to.

7. $7 + \boxed{} = 15$

$15 - 7 = \boxed{}$

8. $5 + \boxed{} = 11$

$11 - 5 = \boxed{}$

9. $\boxed{} + 10 = 20$

$20 - 10 = \boxed{}$

10. $\boxed{} + 9 = 16$

$16 - 9 = \boxed{}$

11. $\boxed{} = 9 + 9$

$9 = \boxed{} - 9$

12. $\boxed{} = 5 + 8$

$5 = \boxed{} - 8$

13. **THINK SMARTER** Solve.
Rick has 10 party hats.
He needs 19 hats for his
party. How many more
party hats does Rick need?

_____ party hats

Math
on the
Spot

Problem Solving • Applications

WRITE Math

Use cubes or draw a picture to solve.

14. Todd has 12 bunnies. He gives 4 bunnies to his sister. How many bunnies does Todd have now?

_____ bunnies

15. Brad has 11 trucks. Some are small trucks. 4 are big trucks. How many small trucks does he have?

_____ small trucks

16. **GO DEEPER** There are 15 children at the park. 6 of the children go home. Then 4 more children come to the park. How many children are in the park now?

_____ children

17. **THINK SMARTER** Use 🔲 🔲 to find the unknown numbers. Write the numbers.

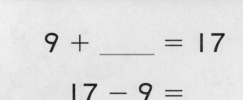

$$9 + \underline{\quad} = 17$$

$$17 - 9 = \underline{\quad}$$

TAKE HOME ACTIVITY • Have your child explain how using subtraction can help him or her find the unknown number in $7 + \square = 16$.

Algebra • Unknown Numbers

Common Core
COMMON CORE STANDARD—1.OA.D.8
Work with addition and subtraction equations.

Write the missing numbers.
Use 🎲 🎲 if you need to.

1. $6 + \boxed{} = 13$

   $13 - 6 = \boxed{}$

2. $9 + \boxed{} = 14$

   $14 - 9 = \boxed{}$

3. $\boxed{} + 7 = 15$

   $15 - 7 = \boxed{}$

4. $\boxed{} = 8 + 8$

   $8 = \boxed{} - 8$

Problem Solving *Real World*

Use cubes or draw a picture to solve.

5. Sally has 9 toy trucks.
   She gets 3 more toy trucks.
   How many toy trucks does
   she have now? _____ toy trucks

6. **WRITE Math** Use words,
   pictures, or numbers to
   show how to find the
   unknown numbers for
   $8 + \underline{} = 17$ and
   $17 - 8 = \underline{}$.

Lesson Check (1.OA.D.8)

I. Write the unknown number.

$$9 + \boxed{} = 16$$

Spiral Review (1.OA.B.3, 1.OA.C.6)

2. What is $14 - 6$?

 Step 1

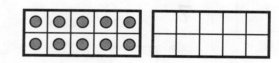 Step 2

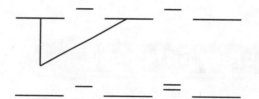

$$\rule{1cm}{0.5pt} - \rule{1cm}{0.5pt} = \rule{1cm}{0.5pt}$$

So, $14 - 6 = \rule{1cm}{0.5pt}$.

3. Draw circles to show the number.
Write the sum. $0 + 8$

$$0 + 8 = \rule{1cm}{0.5pt}$$

FOR MORE PRACTICE
GO TO THE
Personal Math Trainer

Name _____

Algebra • Use Related Facts

Essential Question How can you use
a related fact to find an unknown number?

Common Core **Operations and Algebraic Thinking—1.OA.D.8** *Also 1.OA.C.6*
MATHEMATICAL PRACTICES
MP2, MP4, MP7

Listen and Draw (Real World)

What number can you add to 8 to get 10?
Draw a picture to solve. Write the unknown number.

$$8 + \boxed{} = 10$$

Math Talk

MATHEMATICAL PRACTICES 4

Model Describe how
to solve this problem
using cubes.

FOR THE TEACHER • Have children draw a
picture and complete the number sentence to
show the number that can be added to 8 to
get 10.

Chapter 5

Model and Draw

You can use an addition fact to find a related subtraction fact.

> I know that
> $3 + 7 = 10$, so
> $10 - 3 = 7$.

Find $10 - 3$.

$3 + \underline{\;7\;} = 10$

$10 - 3 = \underline{\;7\;}$

Share and Show MATH BOARD

Write the unknown numbers.

1. Find $14 - 8$.

$8 + \underline{\hspace{1cm}} = 14$

$14 - 8 = \underline{\hspace{1cm}}$

2. Find $17 - 8$.

$8 + \underline{\hspace{1cm}} = 17$

$17 - 8 = \underline{\hspace{1cm}}$

3. Find $11 - 6$.

$6 + \underline{\hspace{1cm}} = 11$

$11 - 6 = \underline{\hspace{1cm}}$

4. Find $15 - 9$.

$9 + \underline{\hspace{1cm}} = 15$

$15 - 9 = \underline{\hspace{1cm}}$

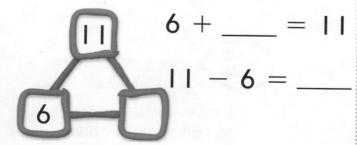
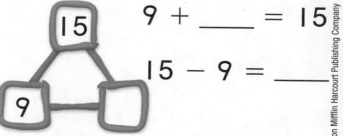

Name _____

Write the unknown numbers.

5. Find 20 − 10.

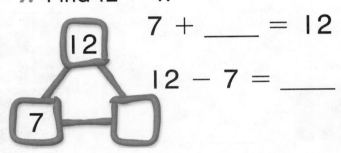

10 + ____ = 20

20 − 10 = ____

6. Find 13 − 4.

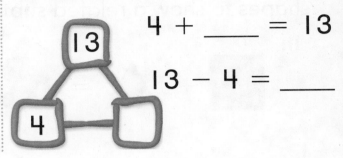

4 + ____ = 13

13 − 4 = ____

7. Find 12 − 7.

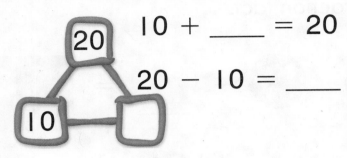

7 + ____ = 12

12 − 7 = ____

8. Find 15 − 8.

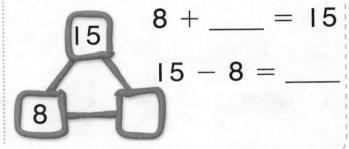

8 + ____ = 15

15 − 8 = ____

GO DEEPER Write an addition sentence to help you find the difference. Then write the related subtraction sentence to solve.

9. Find 11 − 5.

____ + ____ = ____

____ − ____ = ____

10. Find 13 − 6.

____ = ____ + ____

____ = ____ − ____

Problem Solving • Applications  WRITE Math

MATHEMATICAL PRACTICE ② **Reason Abstractly** Look at the shapes in the addition sentence. Draw shapes to show a related subtraction fact.

11.

12.

13. THINK SMARTER

14. THINK SMARTER Which is the unknown number in these related facts?

$$\square + 5 = 12 \qquad 12 - 5 = \square$$

$$5 + \square = 12 \qquad 12 - \square = 5$$

5 ○ 7 ○ 8 ○ 9 ○

 TAKE HOME ACTIVITY • Give your child 5 small objects, such as paper clips. Then ask your child how many more objects he or she would need to have 12.

Algebra • Use Related Facts

Common Core   COMMON CORE STANDARD—1.OA.D.8
Work with addition and subtraction equations.

Write the missing numbers.

1. Find 16 − 9.

    9 + ☐ = 16

    16 − 9 = ☐

    16
    9   ☐

2. Find 12 − 7.

    7 + ☐ = 12

    12 − 7 = ☐

    12
    7   ☐

3. Find 15 − 6.

    6 + ☐ = 15

    15 − 6 = ☐

    15
    6   ☐

4. Find 18 − 9.

    9 + ☐ = 18

    18 − 9 = ☐

    18
    9   ☐

Problem Solving · Real World

Look at the shapes in the addition sentence.
Draw a shape to show a related subtraction fact.

5.

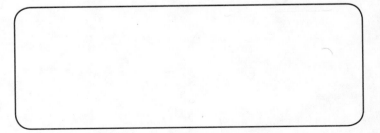

6. **WRITE** Math Draw pictures
    to show how to solve
    14 − 7 = ___ and
    7 + ___ = 14.

Lesson Check (1.OA.D.8)

1. Write an addition fact that helps you solve $12 - 4$.

___ + ___ = ___

Spiral Review (1.OA.C.5, 1.OA.C.6)

2. Circle the greater addend.
Count on to find the sum.

$$\begin{array}{r} 9 \\ + 3 \\ \hline \end{array}$$

3. Draw ⬛ to show the doubles fact.
Write the sum.

$$\begin{array}{r} 8 \\ + 8 \\ \hline \end{array}$$

FOR MORE PRACTICE
GO TO THE
Personal Math Trainer

Name _____

Choose an Operation

Essential Question How do you choose when to add and when to subtract to solve a problem?

Common Core **Operations and Algebraic Thinking—1.OA.A.1** *Also 1.OA.C.6*

MATHEMATICAL PRACTICES
MP3, MP4, MP6

Listen and Draw *Real World* | *Hands On*

Listen to the problem. Use ● to solve.
Draw a picture to show your work.

_____ white balloons

FOR THE TEACHER • Read the following problem. Kira has 16 balloons. She has 8 pink balloons. The other balloons are white. How many white balloons does she have?

Math Talk MATHEMATICAL PRACTICES 6

How did you solve this problem? **Explain.**

Chapter 5

Model and Draw

Mary sees 8 squirrels. Jack sees 9 more squirrels than Mary. How many squirrels does Jack see?

Do you add or subtract to solve?

Explain how you chose to solve the problem.

(add)   subtract   ___ ◯ ___ ◯ ___

___ squirrels

Share and Show   MATH BOARD

Circle **add** or **subtract**.
Write a number sentence to solve.

1. Hanna has 5 markers. Owen has 9 more markers than Hanna. How many markers does Owen have?

   add   subtract

   ___ ◯ ___ ◯ ___

   ___ markers

2. Angel has 13 apples. He gives some away. Then there were 5 apples. How many apples does he give away?

   add   subtract

   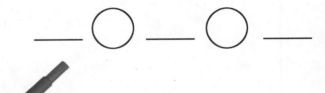

   ___ ◯ ___ ◯ ___

   ___ apples

3. Deon has 18 blocks. He builds a house with 9 of the blocks. How many blocks does Deon have now?

   add   subtract

   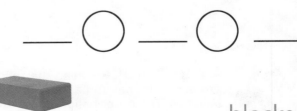

   ___ ◯ ___ ◯ ___

   ___ blocks

Name _____

On Your Own

Circle **add** or **subtract**.
Write a number sentence to solve.

4. Rob sees 5 raccoons. Talia
sees 4 more raccoons than
Rob. How many raccoons
do they see?

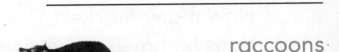

____ raccoons

add   **subtract**

5. Eli has a box with 12 eggs.
His other box has no eggs.
How many eggs are in
both boxes?

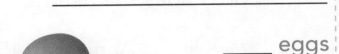

____ eggs

add   **subtract**

6. Leah has a bowl with 16 fish.
Some fish have long tails.
7 fish have short tails. How
many fish have long tails?

____ fish

add   **subtract**

7. **Go DEEPER** Sasha has 8 red
apples. She has 3 fewer
green apples than red
apples. How many apples
does she have?

____ apples

add   **subtract**

Problem Solving • Applications

 WRITE Math

MATHEMATICAL PRACTICE 3 **Apply** Choose a way to solve. Write or draw to explain.

8. James has 4 big markers and 7 skinny markers. How many markers does he have?

_____ markers

9. Sam has 9 baseball cards. She wants to have 17 cards. How many more cards does she need?

_____ more cards

10. **THINK SMARTER** Annie gets 15 pennies on Monday. She gets 1 more penny each day. How many pennies does she have on Friday?

_____ pennies

11. **THINK SMARTER** Beth has 5 grapes. A friend gives her 8 grapes. How many grapes does Beth have now? Draw a picture to show your work.

Beth has _____ grapes.

 TAKE HOME ACTIVITY • Ask your child to write a number sentence that could be used to solve Exercise 9.

Choose an Operation

Common Core **COMMON CORE STANDARD—1.OA.A.1**
*Represent and solve problems involving
addition and subtraction.*

Circle add or subtract.
Write a number sentence to solve.

I. Adam has a bag of II pretzels. He eats
2 of the pretzels. How many pretzels
are left?

    add   subtract

_____

____ pretzels

Problem Solving Real World

Choose a way to solve.
Write or draw to explain.

2. Greg has II shirts.
3 have long sleeves.
The rest have short sleeves.
How many short-sleeve
shirts does Greg have?

____ short-sleeve shirts

3. **WRITE** Math Use words,
numbers, or pictures to
explain another way you
could solve Exercise 2.

Lesson Check (1.OA.A.1)

1. Circle add or subtract. Write a number sentence to solve. There are 18 children on the bus. Then 9 children get off. How many children are left on the bus?

add   subtract

____ ◯ ____ = ____

Spiral Review (1.OA.A.1, 1.OA.B.3)

2. Choose a way to solve. Draw or write to explain. Mike has 13 plants. He gives some away. He has 4 left. How many plants does he give away?

____ plants

3. Write the numbers 3, 2, and 8 in an addition sentence. Show two more ways to find the sum.

____ + ____ + ____ = ____

____ + ____ = ____

____ + ____ = ____

FOR MORE PRACTICE
GO TO THE
Personal Math Trainer

Name _____

Algebra • Ways to Make Numbers to 20

Essential Question How can you add and subtract in different ways to make the same number?

Common Core **Operations and Algebraic Thinking—1.OA.C.6**
MATHEMATICAL PRACTICES
MP5, MP7

Listen and Draw

Use . Show two ways to make 10.
Draw to show your work.

Way One	Way Two

FOR THE TEACHER • Have children use connecting cubes to show two ways to make 10. Then have them draw to show the two ways.

Math Talk

MATHEMATICAL PRACTICES 5

Use Tools How do your models show two ways to make 10?

Model and Draw

How can you make the number 12 in different ways?

You can add or subtract to make 12.

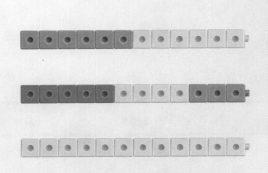

12
_6_ + _6_
_5_ + _4_ + _3_
_12_ – _0_

Share and Show MATH BOARD

Use . Write ways to make the number at the top.

☑ 1.

13
___ + ___
___ – ___
___ + ___ + ___
___ + ___
___ ◯ ___

☑ 2.

10
___ – ___
___ + ___
___ – ___
___ + ___ + ___
___ ◯ ___

Name _____

MATHEMATICAL PRACTICE 5 Use Appropriate Tools

Use ▪ ▪ ▪. Write ways to make the number at the top.

3.

17
___ + ___ + ___
___ + ___
___ – ___
___ ◯ ___

4.

14
___ + ___
___ + ___ + ___
___ – ___
___ ◯ ___

5.

16
___ + ___
___ + ___ + ___
___ – ___
___ ◯ ___

6.

18
___ + ___
___ + ___
___ + ___ + ___
___ ◯ ___

THINK SMARTER Choose a number less than 20. Write the number. Write two ways to make your number. □

7.

8.

Problem Solving • Applications

 Math

GO DEEPER Write numbers to make each line have the same sum.

9.

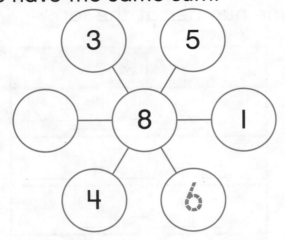

10.

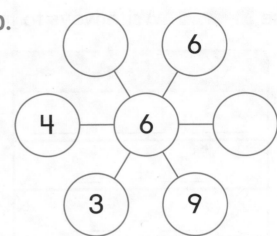

11. **GO DEEPER** Choose a number from 14 to 20 to be the sum. Write numbers to make each line have your sum.

sum for each line ☐

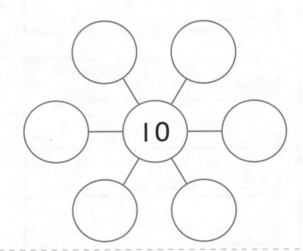

12. **THINK SMARTER** Choose all the ways that make 13.

- ○ 10 + 3
- ○ 9 + 3 + 1
- ○ 8 + 2 + 2

 TAKE HOME ACTIVITY • Have your child explain three different ways to make 15. Encourage him or her to use addition and subtraction, including addition of three numbers.

Algebra • Ways to Make Numbers to 20

 COMMON CORE STANDARD—1.OA.C.6
Add and subtract within 20.

Use . Write ways to make the number at the top.

1. 10

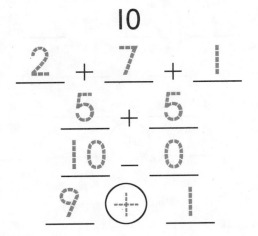

$2 + 7 + 1$

$5 + 5$

$10 - 0$

$9 \oplus 1$

2. 13

___ + ___ + ___

___ + ___

___ − ___

___ $\bigcirc$ ___

Problem Solving Real World

3. Write numbers to make each line have the same sum.

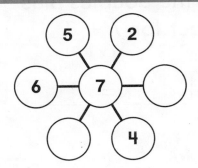

4. WRITE Math Use numbers and pictures to show two ways to make the number 12.

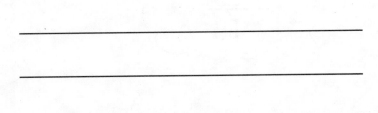

_____

_____

_____

Lesson Check (1.OA.C.6)

1. Write ways to make 18.

18

___ + ___ + ___

___ + ___

___ − ___

___ (+) ___

2. Write ways to make 9.

9

___ + ___ + ___

___ + ___

___ − ___

___ (+) ___

Spiral Review (1.OA.B.4, 1.OA.C.6)

3. Write the doubles plus one fact for 7 + 7.

___ + ___ = ___

4. Write the doubles minus one fact for 4 + 4.

___ + ___ = ___

5. Think of an addition fact to help you subtract.

$$\begin{array}{r} 14 \\ -\ 9 \\ \hline \end{array}$$

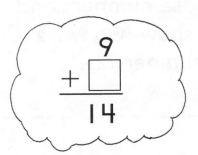

FOR MORE PRACTICE
GO TO THE
Personal Math Trainer

Name _____

Algebra • Equal and Not Equal

Essential Question How can you decide if a number sentence is true or false?

Common Core — Operations and Algebraic Thinking—1.OA.D.7 Also 1.OA.C.6

MATHEMATICAL PRACTICES
MP2, MP6, MP7

Listen and Draw

Color the cards that make the same number.

$2 + 6$	$12 - 6$	$6 + 1$
$13 - 6$	$3 + 3 + 1$	$10 + 6$
$3 + 4$	$4 + 3$	$5 + 2 + 5$
$3 + 2 + 2$	$11 - 2$	$16 - 9$

Math Talk

MATHEMATICAL PRACTICES 2

Reasoning Why can you use two of the cards you color and an equal sign to make a number sentence?

FOR THE TEACHER • Have children color the cards that make the same number.

The equal sign means that both sides are the same.

Write a number to make each true.

4 + 5 = 5 + 5 is **not** true. It is false.

$9 = \underline{9}$ $4 + 5 = \underline{}$ $4 + 5 = \underline{} + 4$

Share and Show MATH BOARD

Which is true? Circle your answer.
Which is false? Cross out your answer.

THINK
Are both sides equal?

1.

$7 = 8 - 1$

~~$1 + 2 = 3 = 2$~~

2.
$4 + 1 = 5 + 2$

$6 - 6 = 7 - 7$

3.
$7 + 2 = 6 + 3$

$8 - 2 = 6 + 4$

4.
$5 - 4 = 4 - 3$

$10 = 1 + 0$

Name _____

MATHEMATICAL PRACTICE 6   **Attend to Precision**

Which are true? Circle your answers.
Which are false? Cross out your answers.

5.

$1 + 9 = 9 - 1$ $8 + 1 = 2 + 7$ $19 = 19$

6.

$9 + 7 = 16$ $16 - 9 = 9 + 7$ $9 - 7 = 7 + 9$

7. THINK SMARTER   Lyle writes the false
number sentence $2 + 10 = 8$.
Complete the number sentence
to make the sentence true.

$2 + \underline{\quad} = 8$

Write numbers to make sentences that are true.

8.

$2 + 10 = 7 + \underline{\quad}$

9.

$\underline{\quad} = 2 + 3 + 4$

10.

$0 + 9 = \underline{\quad} - 9$

11.

$\underline{\quad} + 7 = 7 + 6$

12. THINK SMARTER   Write numbers to show expressions
of equal value.

$\underline{\quad} + \underline{\quad} = \underline{\quad} + \underline{\quad}$

Problem Solving • Applications   WRITE Math

13. Which are true? Use to color.

20 = 20	9 + 1 + 1 = 11	8 − 0 = 8
12 = 1 + 2	10 + 1 = 1 + 10	7 = 14 + 7
	6 = 2 + 2 + 2	
	11 − 5 = 1 + 5	
	1 + 2 + 3 = 4 + 5	

14. **THINK SMARTER** Use the same numbers.
Write a different number sentence
that is true.

$7 + 8 = 15$

___ = ___ ◯ ___

Personal Math Trainer

15. **THINK SMARTER +** Is the math sentence true?
Choose Yes or No.

$5 − 4 = 9 − 8$	○ Yes	○ No
$13 = 5 + 7$	○ Yes	○ No
$6 + 2 = 2 + 6$	○ Yes	○ No

 TAKE HOME ACTIVITY • Write $10 = 7 − 3$ and $10 = 7 + 3$
on a sheet of paper. Ask your child to explain which is true.

Algebra • Equal and Not Equal

Common Core

COMMON CORE STANDARD—1.OA.D.7
Work with addition and subtraction equations.

Which are true? Circle your answers.
Which are false? Cross out your answers.

1. $6 + 4 = 5 + 5$
2. $10 = 6 - 4$
3. $8 + 8 = 16 - 8$

4. $14 = 1 + 4$
5. $8 - 0 = 12 - 4$
6. $17 = 9 + 8$

Problem Solving · Real World

7. Which are true? Use a to color.

$15 = 15$	$12 = 2$	$3 = 8 - 5$
$15 = 1 + 5$	$9 + 2 = 2 + 9$	$9 + 2 = 14$
$1 + 2 + 3 = 3 + 3$	$5 - 3 = 5 + 3$	$13 = 8 + 5$

8. **WRITE** Math Write $5 + \square = 6 + 8$.
Write a number to make the
sentence true. Draw a quick
picture to explain.

_____

_____

1. Circle the number sentences that are true.
   Cross out the ones that are false.

   $4 + 3 = 9 - 2$ $4 + 3 = 9 + 2$

   $4 + 3 = 4 - 3$ $4 + 3 = 6 + 1$

- -

Spiral Review (1.OA.A.2, 1.OA.C.6)

2. Use 5, 6 and 11 to write related
   addition and subtraction sentences.

   ___ ⊕ ___ ⊜ ___

   ___ ⊕ ___ ⊜ ___

   ___ ⊖ ___ ⊜ ___

   ___ ⊖ ___ ⊜ ___

- -

3. Solve. Draw or write to show your work.
   Leah has 4 green toys, 5 pink toys,
   and 2 blue toys. How many toys does Leah have?

   _____ toys

FOR MORE PRACTICE
GO TO THE
Personal Math Trainer

Name _____

Facts Practice to 20

Essential Question How can addition and subtraction strategies help you find sums and differences?

Common Core **Operations and Algebraic Thinking—1.OA.C.6**
MATHEMATICAL PRACTICES
MP2, MP6

Listen and Draw

What is 2 + 8?
Use ●. Draw to show a strategy you can use to solve.

2 + 8 = _____

MATHEMATICAL PRACTICES 6

Explain What other strategy could you use to solve the addition fact?

FOR THE TEACHER • Have children use two-color counters to model a strategy to solve the addition fact. Then have them draw a picture to show the strategy they used.

Model and Draw

Sam is reading a story that has 10 pages. He has read 4 pages. How many pages does he have left to read?

THINK
I can use a related addition fact to solve 10 – 4.

What is 10 – 4?

$4 + \boxed{6} = 10$

So, $10 - 4 = \underline{6}$.

Share and Show

MATH BOARD

Add or subtract.

1. $2 + 5 = \underline{\hspace{1cm}}$

2. $9 - 6 = \underline{\hspace{1cm}}$

3. $\underline{\hspace{1cm}} = 9 + 3$

4. $15 - 7 = \underline{\hspace{1cm}}$

5. $3 - 1 = \underline{\hspace{1cm}}$

6. $\underline{\hspace{1cm}} = 2 + 6$

7. $2 + \boxed{} = 11$

8. $10 - \boxed{} = 2$

9. $8 = 8 + \boxed{}$

10. $12 - 9 = \underline{\hspace{1cm}}$

11. $12 - 4 = \underline{\hspace{1cm}}$

12. $\underline{\hspace{1cm}} = 4 + 9$

13. $\boxed{} + 8 = 13$

14. $\boxed{} - 1 = 6$

15. $9 = \boxed{} + 3$

16. $16 - 7 = \underline{\hspace{1cm}}$

✓ 17. $11 - 8 = \underline{\hspace{1cm}}$

✓ 18. $\underline{\hspace{1cm}} = 8 + 7$

Name _____

On Your Own

MATHEMATICAL PRACTICE 6 Attend to Precision

Add or subtract.

19.
$$\begin{array}{r} 6 \\ + 0 \\ \hline \end{array}$$

20.
$$\begin{array}{r} 17 \\ - 8 \\ \hline \end{array}$$

21.
$$\begin{array}{r} 7 \\ + 4 \\ \hline \end{array}$$

22.
$$\begin{array}{r} 9 \\ - 0 \\ \hline \end{array}$$

23.
$$\begin{array}{r} 17 \\ - 9 \\ \hline \end{array}$$

24.
$$\begin{array}{r} 4 \\ + 6 \\ \hline \end{array}$$

25.
$$\begin{array}{r} 7 \\ + \ \square \\ \hline 10 \end{array}$$

26.
$$\begin{array}{r} 8 \\ - \ \square \\ \hline 3 \end{array}$$

27.
$$\begin{array}{r} 8 \\ + \ \square \\ \hline 11 \end{array}$$

28.
$$\begin{array}{r} 8 \\ - \ \square \\ \hline 2 \end{array}$$

29.
$$\begin{array}{r} 10 \\ - \ \square \\ \hline 6 \end{array}$$

30.
$$\begin{array}{r} 9 \\ + \ \square \\ \hline 17 \end{array}$$

31.
$$\begin{array}{r} 6 \\ + 7 \\ \hline \end{array}$$

32.
$$\begin{array}{r} 4 \\ - \ \square \\ \hline 0 \end{array}$$

33.
$$\begin{array}{r} 5 \\ + \ \square \\ \hline 11 \end{array}$$

34.
$$\begin{array}{r} 13 \\ - 6 \\ \hline \end{array}$$

35.
$$\begin{array}{r} 17 \\ - 9 \\ \hline \end{array}$$

36.
$$\begin{array}{r} 8 \\ + \ \square \\ \hline 16 \end{array}$$

37.
$$\begin{array}{r} 10 \\ + 5 \\ \hline \end{array}$$

38.
$$\begin{array}{r} 13 \\ - 3 \\ \hline \end{array}$$

39.
$$\begin{array}{r} 10 \\ + \ \square \\ \hline 13 \end{array}$$

40.
$$\begin{array}{r} 20 \\ - 10 \\ \hline \end{array}$$

41.
$$\begin{array}{r} 10 \\ - \ \square \\ \hline 9 \end{array}$$

42.
$$\begin{array}{r} 9 \\ + \ \square \\ \hline 19 \end{array}$$

43. **THINK SMARTER** Use the clues to write the addition fact. The sum is 14. One addend is 2 more than the other.

$$\begin{array}{r} \square \\ + \ \square \\ \hline \square \end{array}$$

Problem Solving • Applications  **WRITE**) Math

Solve. Write or draw to explain.

44. There are 14 rabbits. Then 7 rabbits hop away. How many rabbits are there now?

_____ rabbits

45. There are 11 dogs at the park. 2 dogs are gray. The rest are brown. How many dogs are brown?

_____ brown dogs

46. **GO DEEPER** Fill in the blanks. Write the addition fact. Solve.

There are _____ ladybugs on a leaf. Then _____ more ladybugs come. How many ladybugs are there now?

_____ ladybugs

47. **THINK SMARTER** Marco has 13 marbles. Lucy has 8 marbles. How many more marbles does Marco have than Lucy? Write or draw to show your work.

_____ more marbles

 TAKE HOME ACTIVITY • Have your child draw a picture to solve 7 + 4. Then have him or her tell a related subtraction fact.

Facts Practice to 20

 Common Core **COMMON CORE STANDARD—1.OA.C.6**
Add and subtract within 20.

Add or subtract.

1.	2.	3.	4.	5.	6.
4 + 9	13 − 6	4 + 5	8 + 7	11 − 6	17 − 8

7.	8.	9.	10.	11.	12.
5 + 7	13 − 5	16 − 9	3 + 8	9 − 8	7 + 6

13.	14.	15.	16.	17.	18.
9 − ☐ 7	6 + ☐ 10	8 − ☐ 3	6 + ☐ 12	0 + ☐ 9	15 − ☐ 6

Problem Solving *Real World*

Solve. Draw or write to explain.

19. Kara has 9 drawings.
She gives 4 away. How many
drawings does Kara have now?

_____ drawings

20. **WRITE** Math Choose two numbers
from 5 to 9. Use your numbers to
write an addition sentence. Draw
a picture to show your work.

_____

1. Add or subtract.

$$
\begin{array}{r} 14 \\ -\ 7 \\ \hline \end{array}
\qquad
\begin{array}{r} 15 \\ -\ 6 \\ \hline \end{array}
\qquad
\begin{array}{r} 8 \\ +\ 7 \\ \hline \end{array}
\qquad
\begin{array}{r} 5 \\ +\ 8 \\ \hline \end{array}
$$

Spiral Review (1.OA.B.3, 1.OA.D.8)

2. What is the missing number?
   Write the missing addend.

   $7 + \boxed{} = 12$

3. Greg knows $7 + 4 = 11$. What other
   addition fact does he know that
   shows the same addends?
   Write the new fact.

   ___ + ___ = ___

FOR MORE PRACTICE
GO TO THE
Personal Math Trainer

Chapter 5 Review/Test

Personal Math Trainer
Online Assessment
and Intervention

1. There are 2 dogs in the park. Some more
dogs come. Now there are 9 in all. How
many dogs come?

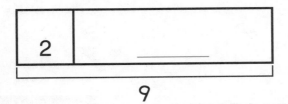

2 dogs   _____ come   9 dogs in all

---

2. Which fact is a related fact?

$$5 + 3 = 8 \qquad\qquad 8 - 5 = 3$$
$$3 + 5 = 8 \qquad\qquad\qquad ?$$

$8 - 3 = 5$ $8 + 5 = 13$ $8 + 3 = 11$ $5 - 3 = 2$
    ○   ○   ○   ○

---

3. Look at the facts. Are they related facts?
Choose Yes or No.

$$14 - 6 = 8 \qquad\qquad 8 + 6 = 14$$

Yes   No

GO DIGITAL Assessment Options
Chapter Test

4. Tom sees 12 bees. Then 7 bees fly away. How many bees does he see now?

Write a number sentence to solve. Then write an addition sentence to check.

5. THINK SMARTER + Use ▣, ▣ to find the unknown numbers. Write the numbers.

$6 + \underline{\hspace{1cm}} = 16$

$16 - 6 = \underline{\hspace{1cm}}$

6. Which is the unknown number in these related facts?

$\boxed{} + 4 = 13$ $\qquad$ $13 - 4 = \boxed{}$

$4 + \boxed{} = 13$ $\qquad$ $13 - \boxed{} = 4$

   5   7   8   9

  ○   ○   ○   ○

7. Joe has 7 blue marbles. A friend gives him 6 red marbles. How many marbles does Joe have now? Draw a picture to show your work.

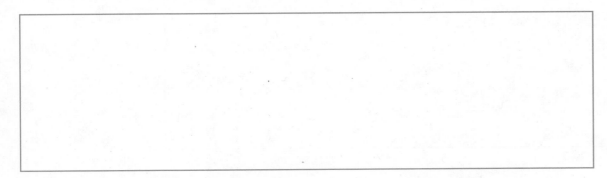

Joe has _____ marbles.

8. Choose all the ways that make 12.

○ $4 + 8$

○ $6 + 5$

○ $5 + 5 + 2$

9. Is the math sentence true? Choose Yes or No.

$7 + 2 = 9 - 2$	○ Yes	○ No
$9 = 6 + 3$	○ Yes	○ No
$5 + 4 = 4 + 5$	○ Yes	○ No

10. Ann has 14 white socks. Bill has 6 white socks. How many more white socks does Ann have than Bill? Write or draw to show your work.

_____

_____ more white socks

11. GO DEEPER   Alma has 5 crayons. Her dad gives her 7 more crayons. How many crayons does she have now? Use a related fact to check your answer.

Alma has _____ crayons.

Write a related fact to check.

_____ − _____ = 5

12. Julia buys 12 books. She gives 9 books away. How many books does she have left?

9	_____

12

_____ books left

© Houghton Mifflin Harcourt Publishing Company